Editor
Betsy Morris, Ph.D.

Editorial Manager
Karen J. Goldfluss, M.S. Ed.

Cover Artist
Tony Carrillo

Art Manager
Kevin Barnes

Creative Director
CJae Froshay

Imaging
James Edward Grace
Rosa C. See

Publisher
Mary D. Smith, M.S. Ed.

How to Work with Data and Probability
Grade 4

BASED ON NCTM STANDARDS

Author
Jennifer Taylor-Cox, Ph.D.

Teacher Created Resources, Inc.
6421 Industry Way
Westminster, CA 92683
www.teachercreated.com

ISBN-1-4206-3740-1

©2005 Teacher Created Resources, Inc.
Made in U.S.A.

The classroom teacher may reproduce copies of materials in this book for classroom use only. The reproduction of any part for an entire school or school system is strictly prohibited. No part of this publication may be transmitted, stored, or recorded in any form without written permission from the publisher.

Table of Contents

How to Use This Book 3
A Note to Teachers and Parents 3
NCTM Standards 4
How to Determine Probability 5
How to Implement the Vocabulary 6

Unit 1: Choose Your Color
How to Conduct the Experiment 7
Choose Your Color Materials 8
How to Calculate the Probability 10
How to Record Outcomes 11

Unit 2: Building Towers
How to Conduct the Experiment 12
How to Record Outcomes 13
How to Draw Conclusions 14
Building Towers Materials 15

Unit 3: In the Bag!
How to Conduct the Experiment 17
How to Record Outcomes 18
How to Draw Conclusions 19

Unit 4: Chips Away!
How to Conduct the Experiment 20
How to Record Outcomes 21
How to Draw Conclusions 22
Chips Away! Materials 23

Unit 5: Digits in a Cup
How to Conduct the Experiment 24
How to Record Outcomes 25
How to Draw Conclusions 26

Unit 6: Fill 'er Up
How to Conduct the Experiment 27
How to Record Outcomes 29
How to Draw Conclusions 30

Unit 7: Spin to 45
How to Conduct the Experiment 31
Spin to 45 Materials 32
How to Record Outcomes 33
How to Draw Conclusions 34

Unit 8: Make a Square
How to Conduct the Experiment 35
Make a Square Materials—Spinners 36
Make a Square Materials—Tangrams 37
How to Record Outcomes 38
How to Draw Conclusions 39

Unit 9: Button Drop for Letters
How to Conduct the Experiment 40
Button Drop for Letters Materials 41
How to Record Outcomes 43
How to Draw Conclusions 44

Answer Key 45

 **Use This Book**

A Note to Teachers and Parents

Welcome to the "How to" math series! You have chosen one of several books designed to give your children the information and practice they need to acquire important concepts in specific areas of mathematics. The goal of the "How to" math books is to give children an extra boost as they work toward mastery of math concepts and skills established by the National Council of Teachers of Mathematics (NCTM).

In this book you will find a host of hands-on probability experiences especially designed for fourth grade students. Each unit experiment includes how to conduct the experiment, how to record the outcomes, and how to draw conclusions related to probability. Each unit provides the children with real life experiences in probability and connected practice in recording outcomes and drawing conclusions.

Each probability experience unit includes "questions to prompt student dialogue." The purpose of using such questions is to encourage children to use math language as they talk about math concepts. Additionally, each unit experience is enhanced with points of differentiation, which can be used to decrease or increase the levels of difficulty for particular individuals or groups of students; thereby allowing all of the probability experiences to be appropriate for students of varying ability and readiness levels.

About This Book

How to Work with Probability: Grade 4 introduces the basic concepts of probability to learners. All of these learners, in their individual backgrounds, have had their own variety of experiences that involved probable outcomes. Some of your students will have had a greater awareness of the laws of probability as they apply to these situations than others. The experiments in this book will help children gain a stronger sense of probability by:

- Building on students' past experiences and supplying new ones
- Asking students questions to help expand their thinking
- Inviting students to reflect upon and explain their thinking
- Giving students opportunities to test and revise their thinking
- Encouraging students to apply logical thinking to everyday experiences
- Helping students separate what they want to happen from what is likely to happen

How to Work with Data and Probability: Grade 4 aligns with the Principles and Standards set forth by the National Council of Teachers of Mathematics (2000).

Probability

The activities in this book help students understand and apply basic concepts of probability. Students discuss the degree of likelihood using such words as certain, equally likely, and impossible. Students predict the probability of outcomes of simple experiments and test their predictions. Students learn to determine the likelihood of an event using words and values.

Data Analysis

The activities in this book include the collection, organization, and display of meaningful data related to probability. Students make predictions based on the data.

Problem Solving

The activities in this book require students to solve mathematics problems in a variety of situations. Students think about how to use and revise problem solving strategies.

Reasoning and Proof

The activities in this book encourage students to make, investigate, and refine mathematics inferences and assumptions. Students support ideas with explanations and justifications.

Communication

The activities in this book require students to use math language to explain their thinking as they answer questions and engage in mathematics dialogue. Students discuss and write about math concepts.

Connections

The activities in this book invite students to connect ideas within mathematics and to connect math ideas to other situations and perspectives. Students make many connections.

Representations

The activities in this book prompt students to record and represent thinking in an organized manner. Students will represent ideas in a variety of ways.

 **Determine Probability**

Facts to Know

Probability is a term in mathematics that means "the chance of something happening." In other words, it is the likelihood of having a specific outcome.

Types of probability include **theoretical** (also called priori) and **experimental** (also called empirical).

Theoretical probability involves the number of specific outcomes and the total number of possible outcomes.

Theoretical probability is what **should** happen.

> Example: If I flip a coin, 50% of the time it should land on heads and 50% of the time it should land on tails. Experimental probability involves the number of times the specific outcome occurs and the number of times you conduct the experiment.

Experimental probability is what **does** happen.

> Example: I flip a coin ten times. Seven times it lands on heads and three times it lands on tails.

Sometimes theoretical probability and experimental probability are aligned. Sometimes they are not aligned. Usually, the more trials in the experiment, the closer the alignment between theoretical probability (what should happen) and experimental probability (what does happen). Therefore, instead of ten flips of the coin, we should try 100 flips of the coin to perform an experiment more representative of what should happen. You could also have five groups of students flip the coin 20 times each and combine the data. Typically, the data from 100 trials in an experiment are more likely to align with theoretical probability than are the data from 10 trials of an experiment. Continuing along this line of thought, the data from 1,000 trials of an experiment are even more likely to align with theoretical probability.

Expressed numerically, probability quantifies the likelihood of some event or outcome. Some future events are highly predictable, and we can assign a higher probability (expressed quantitatively) to these events. To find the probability for a specific action, the possible outcomes need to be known. The likelihood of something happening can be expressed as a number between 0 (impossible) and 1 (certain), or 0% and 100%. For instance, the chance that the sun will set tonight has a 100% probability, because we are certain this will happen. There is only one possible outcome. The chance that you or I will sprout feathers has a 0% probability, because we are certain this will *not* happen. Again, there is only one possible outcome.

© Teacher Created Resources, Inc.

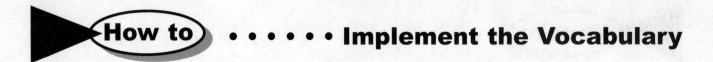

As the teacher or parent, you are encouraged to model the probability vocabulary, inviting students to use these words as they discuss, describe, predict, test and come to understand probability.

Facts to Know

- **Outcome** – result of the experiment or experience

- **Prediction** – anticipated outcome(s) based on data and theory

- **Chance** – the possibility of specific outcomes happening

- **Certain** – outcome will definitely happen; a probability of 100% or 1

    Example: February will immediately follow January

- **Impossible** – outcome will definitely not happen; a probability of 0% or 0

    Example: Alarm clocks growing on trees

- **Likelihood** – the probability or chance of a specific outcome

- **Likely or Probable**—a specific outcome has a greater chance of happening

- **Unlikely or Improbable**—a specific outcome has a smaller chance of happening

- **Equally likely**—the outcomes have equal chances to happen; a probability of 50%

    Example: The toss of a coin has only two possible outcomes, both of which are equally likely to occur. The probability of the coin coming up heads is just as likely as it coming up tails. Both outcomes have a 50/50 chance of occurring. Either outcome expressed as a percent is 50%; expressed as a decimal is .5; and expressed as a fraction is 1/2.

- **Highly likely or More likely**—a specific outcome has a greater chance of happening; a probability of 51% or more

- **Highly unlikely or Less likely**—a specific outcome has lesser chance of happening; a probability of 49% or less

- **Theoretical probability**—the number of specific outcomes divided by the number of possible outcomes.

- **Experimental probability**—the number of times the specific outcome occurs divided by the number of times you conduct the experiment.

#3740 How to ... Data and Probability: Grade 4 © Teacher Created Resources, Inc.

1 Experiment • • • • • • • • • • • • Choose Your Color

Big Idea:

What is the probability of the arrow on a spinner landing on a specific color when the spinner is divided into equal parts?

Preparation: In this probability experiment students spin the arrow on a four-color spinner. Students will discuss the likelihood of the arrow stopping on a specific color. You will need a four-colored spinner (see directions on page 8).

Important Math Concepts:

Equal Chance or Fair Chance means that all of the outcomes have an equal probability of happening.

Directions:

1. Students should sit in a circle and designate someone to record the outcomes.

2. The spinner should pass from one student to the next, with each student having an opportunity to spin the arrow. The recorder should note on the recording sheet the color where the spinner stops each time. (It's helpful to have the children call out the color at the conclusion of each spin.)

3. Repeat, recording the color of each spin. Stop after each student has had a chance to spin the spinner.

4. Prompt the students to discuss the data collected and make predictions about the next round of spins.

5. Continue around the circle, but this time ask each student to predict what color they think the arrow will land on before they spin.

Questions to Prompt Student Dialogue:

- When you spin the arrow, on which color do you think it will stop?
- If we spin the arrow twelve times, how many times do you think it will land on one specific color (example: red)?
- If the arrow lands on red on the first spin, how likely is it that it will land on red on the next spin?
- How likely is it that the arrow will come to rest on a different color on the next spin?
- Are any outcomes certain?

Points of Differentiation:

To decrease the level of difficulty, use a spinner divided in half with only two colors.

To increase the level of difficulty, use a spinner with more colors (see page 8).

© Teacher Created Resources, Inc.

Choose Your Color

How to Make a Spinner

Commercial spinners from a teacher supply store or a board game work best. However, you can make your own spinner this way:

- Take a 6" by 6" (16cm x 16cm) piece of cardboard or tagboard. Draw a circle on the cardboard using the pattern on page 9.
- Use a heavy marker to divide the circle into 4 equal parts.
- Color each section a different color (blue, green, yellow, red).
- Make an arrow out of another piece of cardboard using the pattern on page 9. (The length of the arrow should be no greater than the radius of the circle.)
- Fasten the arrow to the middle of the cardboard circle using a brad or paper fastener.
- Or, simply hold a paperclip in place in the center of the circle using the point of a pencil.

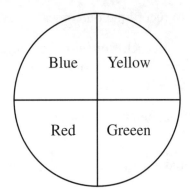

For this experiment, the spinner should be divided into four equal parts as shown:

To simplify the experiment, reduce the number of possible outcomes. Divide the spinner into two equal sections.

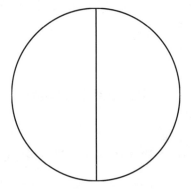

To increase the difficulty level of the experiment, increase the number of possible outcomes, or make the chance of any certain outcome more or less likely by dividing the spinner into unequal sections.

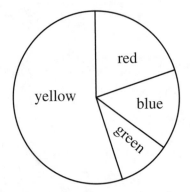
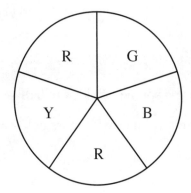
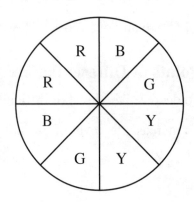

#3740 How to ... Data and Probability: Grade 4   © Teacher Created Resources, Inc.

1 Materials ············ Choose Your Color

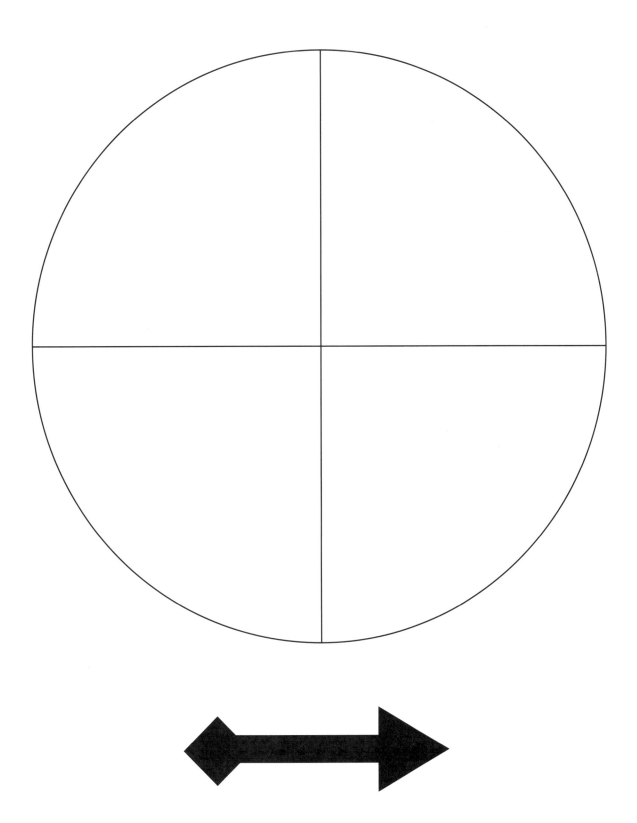

1 ▸ How to · · · · · · · Calculate the Probability

Probability as a Fraction

It is important for students to realize that the probability of certain outcomes occurring can be expressed quantitatively. The likelihood of an event can be shown as a **fraction**, a **decimal**, or a **percent.**

One way to show probability (or the likelihood of an event) is as a **fraction**.

The numerator of the fraction is the number of chances for the event to occur.

In the case of a spinner with four sections and each section being a different color, what is the chance of the arrow landing on a specific color (red, for instance)? There is only one opportunity for the spinner to land on the red section, because only one section of the wheel is red. So the numerator of the fraction is "1."

The denominator of the fraction is the total number of possible outcomes.

In the case of this spinner, there are only four possible outcomes: red, blue, green and yellow. So the denominator of the fraction is "4."

Expressed as a fraction, the probability or chance of the arrow landing in the red section is:

$$\frac{\text{Chance of a specific outcome (red)}}{\text{All possible outcomes of an event}} \qquad \frac{1}{4}$$

Probability as a Decimal

Another way to show probability is with a decimal. Divide the numerator by the denominator. Using the above example:

$$\frac{\text{Chance of a specific outcome (red)}}{\text{All possible outcomes of an event}} \qquad \frac{1}{4} = .25$$

Probability as a Percent

A third way to show probability is with a **percent**. Multiplying the decimal expression by 100 will yield that quantity as a percent. Using the same example:

.25 X 100 = 25 (with the % symbol) or 25%

So the probability of the arrow landing on red is 25%

#3740 How to ... Data and Probability: Grade 4   10   © *Teacher Created Resources, Inc.*

1 How to • • • • • • • • • • • • Record Outcomes

Tally Marks: A way of counting items using lines (1's) and bundles of lines (5's)

| = 1 || = 2 ||| = 3 |||| = 4   ЖЖ = 5   ЖЖ ЖЖ = 10

ЖЖ ЖЖ | = 11   ЖЖ ЖЖ || = 12   ЖЖ ЖЖ ||| = 13

Use tally marks to complete the table.

Choose Your Color

	Red	Green	Yellow	Blue
Total				

Color the sections to complete the Bar Graph.

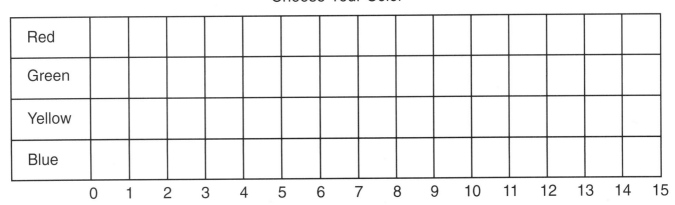

Choose Your Color

Drawing Conclusions

Describe the outcomes of your probability experience. _____

How are the outcomes related to what you predicted? _____

Describe a change that could be made in this experiment that would make the outcome more predictable. For instance, what could you do to increase the chances of the arrow landing on a particular color?

_____

_____

_____

What could you do that would make the outcome harder to predict? Explain your answer.

_____

_____

2 Experiment • • • • • • • • • • • • • • Building Towers

Big Idea:
What is the probability of the arrow on a spinner landing on a specific color if the areas of the spinner are unevenly divided?

Important Math Concepts:
A *prediction* is the anticipated outcome. A prediction should be based on theory and data.

Certain means that outcome will definitely happen.

Impossible means the outcome will definitely not happen.

Preparation:
This probability experiment builds on the previous experiment. Students will determine the probability associated with a variety of spinners divided into uneven sections of color.

Students should pair off for this activity. Each team will need linking cubes, one type of spinner (see spinner options on pages 15–16), and a copy of the activity sheet on page 13. Students should examine their spinner and make predictions about the outcomes.

Directions:
1. This activity begins with pairs of students working together to build a tower of twenty cubes.
   Note: This probability experiment can also be conducted using colored chain links. At the end of the data collection, links can be joined into a circle and compared to the spinner.
2. Students take turns spinning the spinner and choosing the matching color cube.
3. Students should start a new "tower" the first time the arrow lands on a color on the spinner, and add to that tower for subsequent spins. In other words, when the arrow lands on the blue section of the spinner, add a blue cube to the blue tower.
4. After spinning twenty times, encourage the individual teams to compare and contrast the height of each colored tower to the corresponding colored section of their spinners.
5. Tell students to note the relationship between the height of the tower (of a particular color) and the size of that color section on the spinner. Students should record their observations.

When all the teams have finished, they should bring their towers to an area of the classroom where the whole class can convene. Each team can display their towers/spinners, and report the correlation they observed between the tower and the spinner.

Questions to Prompt Student Dialogue:
- How likely is it that you will get a specific color (example: green)?
- Are you certain that your data will include all colors?
- Are there any colors that are impossible?
- Are there any colors that are certain?
- Can anyone predict what it might look like if we were to put all the towers together into one class tower? Why?

Points of Differentiation:
To decrease the level of difficulty, use a spinner with fewer sections, or sections divided equally.

To increase the level of difficulty, use a spinner with more sections, or a spinner with unequal sections.

2 How To • • • • • • • • • • • • Record Outcomes

Building Towers

Use tally marks to record the number of times the arrow landed on each color.

	Tally Marks	Total Number
_____ Color		
_____ Color		
_____ Color		
_____ Color		

Place the towers of colored cubes next to one another. Describe what you see. How do the towers look like a bar graph? Draw a representation of what you see on the bar graph below, filling in the columns with the appropriate colors. Label each axis and include a title.

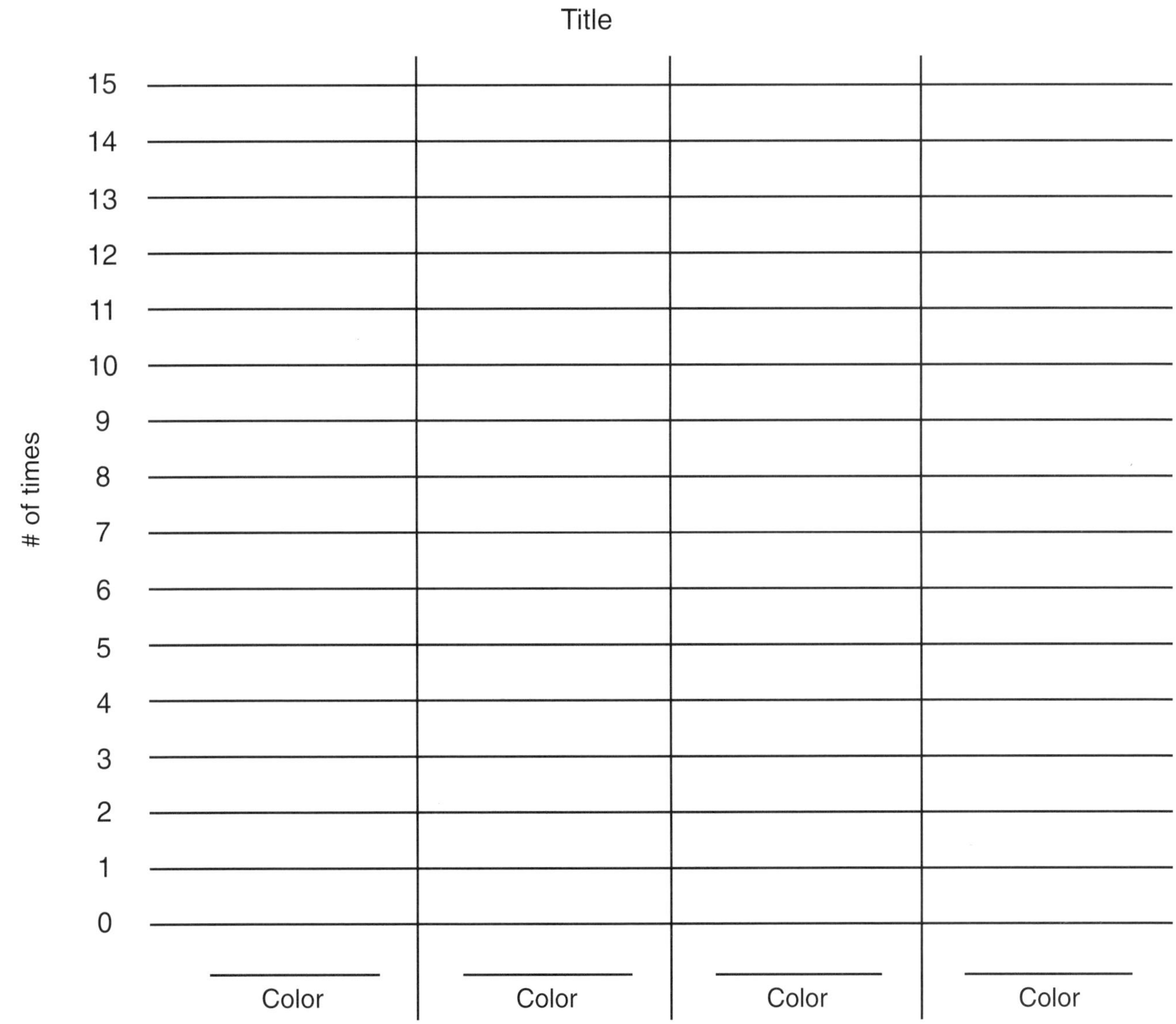

Title

# of times	15
	14
	13
	12
	11
	10
	9
	8
	7
	6
	5
	4
	3
	2
	1
	0

_____ Color   _____ Color   _____ Color   _____ Color

© Teacher Created Resources, Inc.   13   #3740 How to ... Data and Probability: Grade 4

 **Draw Conclusions**

Building Towers

Describe the outcomes of your probability experience.

_____
_____
_____

How are the outcomes connected to what you predicted?

_____
_____
_____

What do you think will happen if you conduct this probability experiment again? Explain your answer.

_____
_____
_____
_____

Describe a change that could be made in this experiment.

_____
_____
_____

Predict the outcomes of the modified probability experiment. Explain your answer.

_____
_____
_____

2 Materials • • • • • • • • • • • • • • Building Tower

Spinner Options (Equal sections)

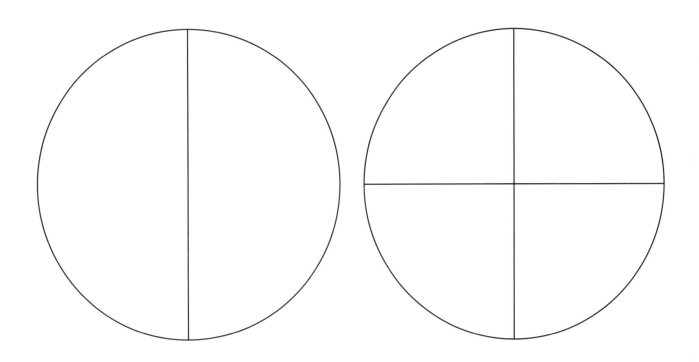
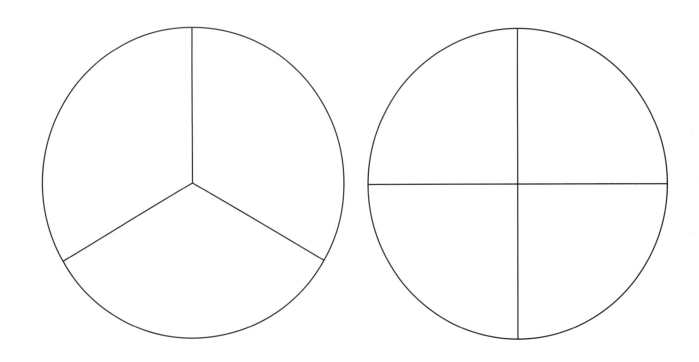

2 Materials • • • • • • • • • • • • • • • Building Tower

Spinner Options (Unequal sections)

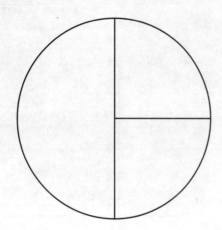
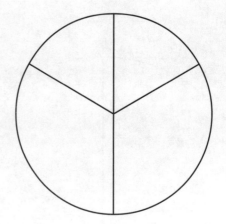
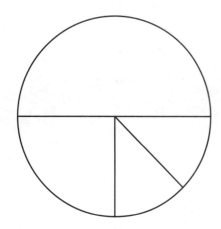
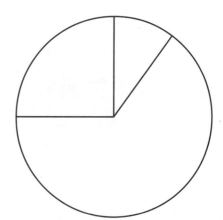
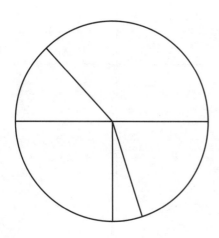
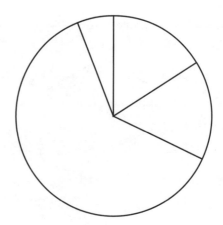

#3740 How to ... Data and Probability: Grade 4 © Teacher Created Resources, Inc.

3 Experiment · · · · · · · · · · · · · · · · · · · In the Bag!

Big Idea:
What is the probability of drawing specific colored crayons from the bag?

Important Math Concepts:
Random selection means choosing something by chance.
Probability can be shown as a fraction or as a percent.

Preparation:
For this probability experiment, you will need an opaque bag, four brown crayons, three purple crayons, two red crayons, and two yellow crayons.

Someone draws four randomly selected crayons from the bag. The colors drawn are recorded for each trial. Comparing the results of each individual draw, students will try to predict which of the following five statements is most likely to be true about the color crayons in the bag.

There are more purple crayons.
There are less red crayons.
There are exactly two brown crayons.
At least one of the crayons is red.
There are no yellow crayons.

Directions:
1. Post the statements where students can refer to them.
2. Discuss the likelihood of each statement being true of the crayons selected when four crayons are drawn randomly from the bag.
3. After the students have had an opportunity to make their predictions, begin drawing four random crayons (without looking in the bag).
4. Using the worksheet on page 18, place a tally mark beside each statement that is true of the four crayons selected.
5. Place the crayons back in the bag, shake them up, and draw again.
6. Continue drawing four crayons at a time and recording the results. Collect data from 100 draws (by combining data sets collected over several days by small groups of students) before discussing conclusions about the probability of each statement.
7. Try this experiment again, at a later time and compare the data sets, describing the related probability.

Questions to Prompt Student Dialogue:
- Which of the statements above are probably true?
- Are there any statements that are equally likely?
- Which statement is unlikely?

Points of Differentiation:
To decrease the level of difficulty use fewer crayons and/or statements.
To increase the level of difficulty change the number of crayons and/or encourage students to create new likelihood statements.

© Teacher Created Resources, Inc.   #3740 How to ... Data and Probability: Grade 4

3 ▶ How to • • • • • • • • • • • • • Record Outcomes

In the Bag!

Use tally marks to complete the table.

_____
Title of Table

Statement	Yes	Yes	Yes	Yes	Yes	Yes	Yes	Yes	Yes	Yes	Total
More purple crayons											
Less red crayons											
Exactly two brown crayons											
At least one red crayon											
No yellow crayons											

Trial 1 Trial 2 Trial 3 Trial 4 Trial 5 Trial 6 Trial 7 Trial 8 Trial 9 Trial 10

Using the data from 100 trials, find the percentage of time each of the statements above was true. For example, if the statement was true 25 out of 100 times, the probability can be shown as a fraction $\frac{25}{100}$ or as a percent 25%.

Statement 1 _____

Statement 2 _____

Statement 3 _____

Statement 4 _____

Statement 5 _____

#3740 How to ... Data and Probability: Grade 4   18   © Teacher Created Resources, Inc.

Draw Conclusions

In the Bag!

Describe the outcomes of your probability experience.

_____

_____

How are the outcomes connected to what you predicted?

_____

_____

What do you think will happen if you conduct this probability experiment again? Explain your answer.

_____

_____

Describe a change that could be made in this experiment.

_____

_____

Predict the outcomes of the modified probability experiment. Explain your answer.

_____

_____

4 Experiment · · · · · · · · · · · · · · · · · · Chips Away!

Big Idea:
What is the probability of getting specific scores using the Chips Away! Game Board?

Important Math Concepts:
Likelihood is the probability or chance of a specific outcome.

Unlikely means a specific outcome has a lesser chance of happening.

Preparation:
Using the Chips Away! Game Board on page 23 and two "chips" (counters or markers), students take turns throwing the chips on the board. Students find the total of the two numbers on which the chips land. This is the "score" of that turn.

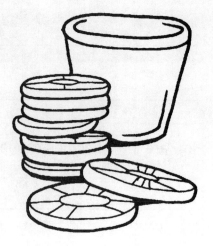

Make predictions before you begin the probability experiment.

Directions:
1. Have one student step forward and toss the two "chips" onto the Chips Away! Game Board.
2. Total the numbers under each chip and record the sum on the worksheet on page 21.
3. Have the next student step forward and toss two chips.
4. Conduct 20 trials and keep a record of total scores.
5. As with other experiments, it is a good idea to combine data sets with other groups of students to increase the number of outcomes to examine.

Questions to Prompt Student Dialogue:
- How likely is it to get a high score?
- What is the probability of getting a score of twenty or less?
- What outcome is highly unlikely in this experiment?
- What outcome is impossible?

Points of Differentiation:
To decrease the level of difficulty, create a game board with fewer, smaller numbers. Counters and number lines can be used to help students tabulate scores.

To increase the level of difficulty, allow the manner in which the chip lands to determine the score (example: up-side-down = double score).

#3740 How to ... Data and Probability: Grade 4   © Teacher Created Resources, Inc.

 • • • • • • • • • • • • • • • • **Game Board**

Chips Away!

Record the sums (one total in each box).

_____
Title of Table

Use an X to indicate each score on the line plot. Be sure to place your x on the intersection.

_____
Title of Line Plot

10
9
8
7
6
5
4
3
2
1st Trial

0 1 2 3 4 5 6 7 8 9 10 11 12 13 14 15 16 17 18 19 20 21 22 23 24 25 26 27 28 29 30 31 32 33 34 35 36 37 38 39 40

© Teacher Created Resources, Inc.   21   #3740 How to ... Data and Probability: Grade 4

 **Draw Conclusions**

Chips Away!

Describe the outcomes of your probability experience.

_____
_____
_____

How are the outcomes connected to what you predicted?

_____
_____
_____

What do you think will happen if you conduct this probability experiment again? Explain your answer.

_____
_____
_____

Describe a change that could be made in this experiment.

_____
_____
_____

Predict the outcomes of the modified probability experiment. Explain your answer.

_____
_____
_____

4 Materials ················ Game Board

Chips Away

5 Experiment • • • • • • • • • • • • • • • Digits in a Cup

Big Idea:

Using only the digits 0 through 9, what is the probability of drawing any specific number?

Important Math Concepts:

Likelihood is the probability or chance of a specific outcome.

More likely means a specific outcome has a greater chance of happening; a probability of 51% or more.

Less likely means a specific outcome has a lesser chance of happening; a probability of 49% or less.

Preparation:

In this probability experiment students place the digits 0 – 9 in a cup. You can either use any number tiles that you have available or write the individual digits on small paper squares.

Directions:

1. Encourage students to discuss the probability associated with this experiment by predicting the digit value before drawing.
2. Students randomly draw one tile (or paper).
3. Students record this number on the worksheet on page 25.
4. Students can record results using tally marks to show how often each digit is drawn.
5. Students should perform at least 20 trials.
6. Students indicate whether the number is more than 5, less than 5, or exactly 5.
7. Encourage conversation regarding the likelihood of outcomes. How is the experiment altered if the tile drawn is not replaced before drawing the next tile?

Questions to Prompt Student Dialogue:

- How likely is it that the number will be more than 5?
- What is the likelihood of drawing exactly 5?
- Is it more likely to draw exactly 5 or less than 5?

Points of Differentiation:

To decrease the level of difficulty, have students draw only digits 0-5 and record more than 3, less than 3, and exactly 3.

To increase the level of difficulty, students can draw two-digit numbers and record values of more than 50, less than 50, or exactly 50.

5 How To ····· Record Outcomes

Digits in a Cup

Use tally marks to complete the table.

_____
Title of Table

More than 5	Less than 5	Exactly 5

Color the sections of the circle graph below to indicate how many trials resulted:

In a number greater than five? _____

Color that many sections of the circle red.

In a number less than five? _____

Color that many sections of the circle blue.

In a number equal to five? _____

Color that many sections of the circle black.

_____
Title of Circle Graph

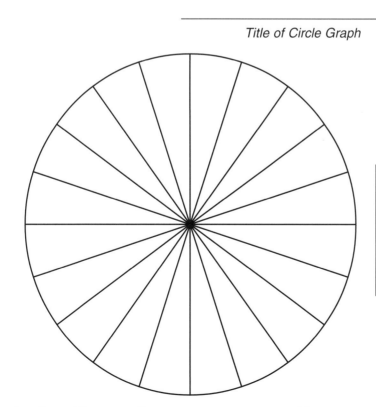

Key:

More than 5:

Less than 5:

Exactly 5:

© Teacher Created Resources, Inc. 25 #3740 How to ... Data and Probability: Grade 4

Draw Conclusions

Digits in a Cup

Describe the outcomes of your probability experience.

_____

_____

How are the outcomes connected to what you predicted?

_____

_____

What do you think will happen if you conduct this probability experiment again? Explain your answer.

_____

_____

Describe a change that could be made in this experiment.

_____

_____

Predict the outcomes of the modified probability experiment. Explain your answer.

_____

_____

6 Experiment •••••••••••••••••• Fill 'Er Up

Big Idea:
Which number cube (die) offers the greater chance of filling the cup in the least number of rolls?

Important Math Concepts:
Greater chance means more likely.

Preparation:

This probability experiment requires three 6-sided number cubes (which can be made by writing numbers on small blocks).

Each cube needs to be identified somehow by either:
- using different colored cubes;
- marking one side of the first cube with the letter A, the second cube with the letter B, and the third cube with the letter C;
- or marking each cube with a symbol such as a star, a fish, and a smiley face.

Label the first number cube with the digits 1, 2, 3, 4, 5, 6.

Label the second number cube with the digits 1, 1, 2, 2, 3, 3. Label the third number cube with the digits 4, 4, 5, 5, 6, 6. You will also need cups, spoons, and rice or water.

Note: Be sure that the same size cup and the same size spoon is used for each trial. You will also want to do a test to be sure that it does not take more than 14 spoonfuls of rice (or water) to fill the cup. If it does, use smaller cups or larger spoons.

This experiment is done most efficiently if the students work in pairs, but may also be done individually.

Directions:

1. Divide the students into pairs. Designate one student as the "roller" and the other as the "recorder." They should take turns rolling the die.
2. The "roller" chooses one of the three number cubes available, and uses this cube for the duration of the first trial. The "recorder" must be sure to note which number cube was chosen for this trial on the worksheet (page 29).
3. The "roller" rolls the cube, and places the corresponding number of spoonfuls of rice (or water) in the cup.
4. The "recorder" uses a tally mark to mark the number rolled.
5. The "roller" continues to roll the number cube and place spoonfuls in the cup until the cup is full.
6. The "recorder" should use tally marks to record the number rolled on each throw.
7. Students record how many rolls it took to fill the cup using the first, second, or third number cube. Students should make predictions before and during the experiment.
8. The pair should change roles and repeat the experiment, this time choosing a different number cube.

As with other probability experiments, the data can be combined to find the average number of rolls it takes with each number cube to fill the cup.

© Teacher Created Resources, Inc.   27   #3740 How to ... Data and Probability: Grade 4

Questions to Prompt Student Dialogue:

- How likely is it that you will get to put in five spoonfuls on your turn?
- How likely is it that you will get to put less than three spoonfuls in the cup?
- Considering the probability associated with your number cube, how many turns do you think it will take to fill up the cup?
- Which number cube should you use if you want to fill the cup in the least number of rolls?

Points of Differentiation:

To decrease the level of difficulty, use only two different number cubes.

To increase the level of difficulty, use more than three different types of number cubes.

Record Outcomes

Fill 'er Up

Use tally marks to complete the table.

Example:

	1s Rolled	2s Rolled	Which Cube Used	Number of Turns
Trial 1	|||	||	☆	5

Title of Table

	1s Rolled	2s Rolled	3s Rolled	4s Rolled	5s Rolled	6s Rolled	Which Cube Used	Number of Turns
Trial 1								
Trial 2								
Trial 3								

Use Xs to plot the number of times each score was rolled. You will need to make a separate graph for each trial.

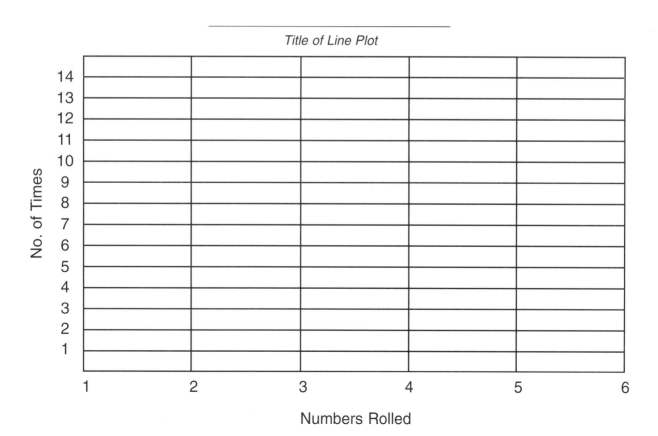

 **Draw Conclusions**

Fill 'er Up

Describe the outcomes of your probability experience.

_____
_____
_____

How are the outcomes connected to what you predicted?

_____
_____
_____

What do you think will happen if you conduct this probability experiment again? Explain your answer.

_____
_____
_____

Describe a change that could be made in this experiment.

_____
_____
_____

Predict the outcomes of the modified probability experiment. Explain your answer.

_____
_____
_____

7 Experiment · · · · · · · · · · · · · · · · · · · Spin to 45

Big Idea:

What is the probability of scoring 45 in a specific number of spins?

Important Math Concepts:

Most likely means a specific outcome has a greater chance of happening.

Least likely means a specific outcome has a lesser chance of happening.

Preparation:

Students add numbers to reach 45 using various spinners. Spinners should be labeled as shown on page 32. Label the back of each spinner with a Letter (A, B, C) so that the students can record which spinner was used for each trial.

Each student needs at least one spinner, and a copy of the recording sheet on page 33.

Directions:

1. Post the spinners. Looking at the first spinner, discuss the likelihood of reaching 45 in 2 spins; in 5 spins; in 10 spins, etc. For each spinner, let students predict the number of spins it would take to total 45.
2. After the discussion, allow students to make predictions about each spinner and record their guesses prior to completing the experiment.
3. Each student chooses a spinner. Working alone, the student needs to record how many spins it takes to reach 45 (or greater). Be sure to remind them to record which spinner was used for each trial.
4. Each student should try at least three spinners.
5. Afterwards, post the outcomes and encourage students to describe the differences in probability associated with the spinners.
6. Order the spinners from most likely (to reach 45 in the least spins) to least likely (to reach 45 in the least spins).

Questions to Prompt Student Dialogue:

- Are you certain that you will reach 45 in less than 45 spins?
- What is the likelihood of reaching 45 in ten spins?
- Which spinner is more likely to help you reach 45 in the least amount of spins?
- Which spinner is most likely to need the most spins to reach 45?

Points of Differentiation:

To decrease the level of difficulty, spin to 20 and use only two types of spinners. Counters and number lines can be used to help students tabulate scores.

To increase the level of difficulty spin to 100, using several types of spinners. This activity can be extended by having students create their own spinners to test predictions about experimental and theoretical probability.

© Teacher Created Resources, Inc.

# Spinners

Spin to 45

Spinner Options

1 2 3 4	5 6 7 8
2 4 6 8	1 3 5 1
0 1 2 9	1 0 0 10

#3740 How to ... Data and Probability: Grade 4   32   © Teacher Created Resources, Inc.

Record Outcomes

Spin to 45

Use tally marks to record how many spins it takes to reach 45.

Title of Table

Spinner:	Spinner:	Spinner:

Fill in or mark the sections to complete the bar graph.

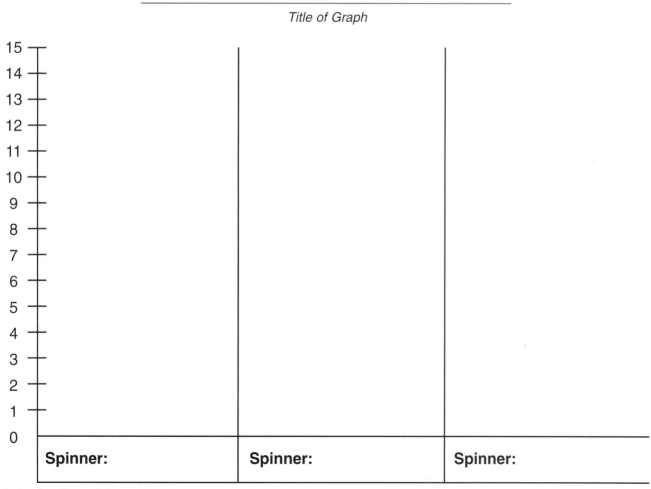

 •••••••••••• **Draw Conclusions**

Spin to 45

Describe the outcomes of your probability experience.

_____
_____
_____

How are the outcomes connected to what you predicted?

_____
_____
_____

What do you think will happen if you conduct this probability experiment again? Explain your answer.

_____
_____
_____

Describe a change that could be made in this experiment.

_____
_____
_____

Predict the outcomes of the modified probability experiment. Explain your answer.

_____
_____
_____

8 Experiment •••••••••••••••• Make a Square

Big Idea:
What is the probability of building a tangram square using a specific spinner?

Important Math Concepts:
Likelihood means the probability or chance of a specific outcome.

A triangle has three sides. A parallelogram is a quadrilateral that has two sets of parallel sides.

Preparation:
In this probability experiment, students will explore probability by seeing how many spins it takes to build a tangram square. Assign the students to work in pairs. Each team will need:

A copy of the recording sheet on page 38

Spinners labeled with tangram shapes (page 36)

Cut-out tangram pieces (Either use pieces you have readily available, or have the students create their own cut-outs by coloring in the sections of the tangram square on page 37.)

Directions:
1. Using one of the spinners (see spinner options on page 36), students work in pairs taking turns spinning and building individual tangram squares.
2. If a student spins a piece that they already have, they cannot place/color the piece, and it becomes the next player's turn.
3. Students should keep a record of how many spins it takes to "make a square."
4. Afterwards, display the spinners and the data collected. Teachers should encourage the students to see the relationship between how many spins it takes and the probability associated with each spinner.

Questions to prompt student dialogue:
- What is the likelihood of completing your square in seven spins?
- What is the probability that you will still need a triangle after ten spins?
- Which spinner is most likely to help you build a tangram square in the least number of spins?
- How would you order the spinners according to the probability of each?

Points of Differentiation:
To decrease the level of difficulty, students can use only two types of spinners.

To increase the level of difficulty, students can use all the spinner types and create new spinners. Students can also try making two squares at one time.

© Teacher Created Resources, Inc.   35   #3740 How to ... Data and Probability: Grade 4

Make a Square
Spinner Options

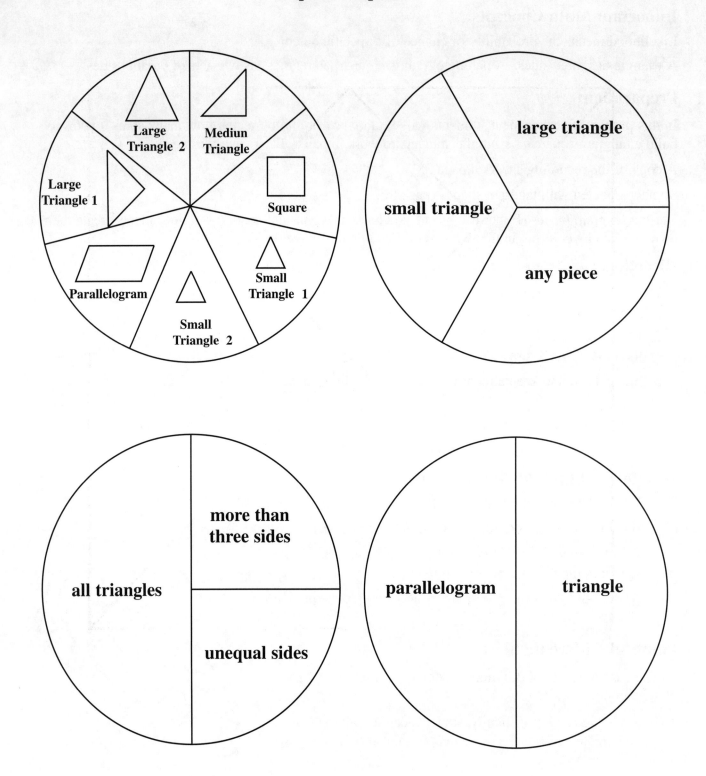

#3740 How to ... Data and Probability: Grade 4

8 Materials · Tangrams

Make a Square

Tangram Pieces

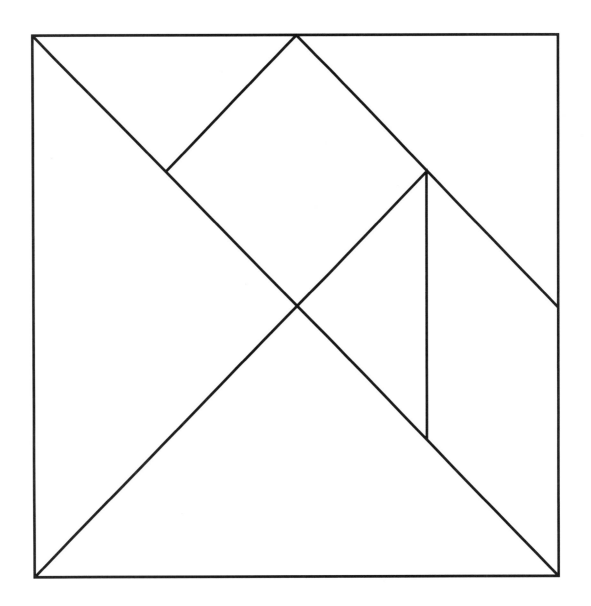

© Teacher Created Resources, Inc.   37   #3740 How to ... Data and Probability: Grade 4

8 ►How to • • • • • • • • • • • • Record Outcomes

Make a Square

Use tally marks to complete the table. Use the frame provided or construct your own.
Remember to add title.

_____

Title of Table

square	small triangle	medium triangle	large triangle			

Using the date above, fill the sections to complete the bar graph. Remember to label each
axis and include a title.

_____

Title of Graph

 **Draw Conclusions**

Make a Square

Describe the outcomes of your probability experience.

_____
_____

How are the outcomes connected to what you predicted?

_____
_____

What do you think will happen if you conduct this probability experiment again? Explain your answer.

_____
_____

Describe a change that could be made in this experiment.

_____
_____

Predict the outcomes of the modified probability experiment. Explain your answer.

_____
_____

Button Drop for Letters

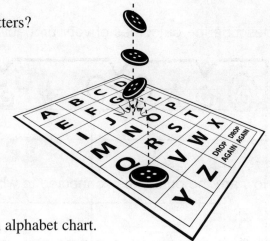

Big Idea:

What is the probability of dropping the button on specific letters?

Important Math Concepts:

To find the theoretical probability, divide the number of specific outcomes by the number of possible outcomes.

To find the experimental probability, divide the number of times the specific outcome occurs by the number of times you conduct the experiment.

Preparation:

In this probability experiment students drop a button onto an alphabet chart.

Directions:

1. Encourage students to predict, discuss, and describe the probability of the button landing on a specific letter.

2. Using the activity sheet on page 43, have students use tally marks to record what letter the button lands on for each drop.

3. The button should be dropped at least 100 times. (If necessary, have students work in pairs to reduce the amount of time this experiment takes. One student can drop the button while the other student records the letter the button landed on. After 50 trials, have the students switch roles.)

4. A letter of the alphabet should then be randomly assigned to each student (or pair of students).

5. Using the sectioned row of the bar graph on page 43, students should fill in the number of sections that correspond to the number of times the button dropped on their assigned letter. For instance, if they were assigned the letter "A," and the button fell on that letter 5 times, they should color in 5 sections of the bar.

6. Students should cut out their bar and add it to a bar graph assembled by the entire class to show how many times the button fell on each letter of the alphabet.

Questions to Prompt Student Dialogue:

- Which letter (if any) is the button most likely to land on?
- Which letters (if any) are less likely? Why?
- What is the probability of the button landing on one of the letters in your name?
- How likely is it for the button to land on a consonant?

Points of Differentiation:

To decrease the level of difficulty students can use one row or column of letters or create their own "Drop for Letters Chart."

To increase the level of difficulty try using the optional alphabet charts and encouraging students to make their own "Drop for Letters Chart."

#3740 How to ... Data and Probability: Grade 4

9 ▶ **Materials** • • • • • • • • • • • • • • • **Game Board**

Button Drop for Letters

A	B	C	D
E	F	G	H
I	J	K	L
M	N	O	P
Q	R	S	T
U	V	W	X
Y	Z	Drop Again	Drop Again

© *Teacher Created Resources, Inc.* 41 *#3740 How to ... Data and Probability: Grade 4*

Game Board

Button Drop for Letters
Optional Chart

m	a	k	b
i	l	e	h
e	o	g	u
i	a	u	j
u	d	a	n
f	o	i	o
e	u	c	u

9 ▶ How to • • • • • • • • • • • • • Record Outcomes

Button Drop for Letters

Use tally marks to show how many times the button landed on each letter.

A	
B	
C	
D	
E	
F	
G	
H	
I	
J	
K	
L	
M	

N	
O	
P	
Q	
R	
S	
T	
U	
V	
W	
X	
Y	
Z	

Now, let's make a large graph with everyone contributing a letter strip. In the first section of the bar below, write the letter that was assigned to you. Fill in the remaining sections of the bar to indicate the number of times the button landed on your assigned letter. Connect your part of the graph to the graph constructed by your classmates.

Using the graph constructed by the whole class, write a fraction for each letter of the alphabet to represent the probability outcome. Write the fractions next to each letter in the chart at the top of this page.

Example: Button landed on Letter ☐ ___3___ times out of 100 trials.

The fraction is $\frac{3}{100}$

© Teacher Created Resources, Inc.   43   #3740 How to ... Data and Probability: Grade 4

 **••••••••• Draw Conclusions**

Button Drop for Letters

Describe the outcomes of your probability experience.

_____

_____

How are the outcomes connected to what you predicted?

_____

_____

What do you think will happen if you conduct this probability experiment again? Explain your answer.

_____

_____

Describe a change that could be made in this experiment.

_____

_____

Predict the outcomes of the modified probability experiment. Explain your answer.

_____

_____

Answer Key

Unit 1: Choose Your Color Pg. 11

Describe the outcomes of your probability experience.

Answers should include how many times the arrow landed on each color of the spinner. For example, 25 blue, 25 red, 25 yellow, 25 green

How are the outcomes connected to what you predicted?

Answers should include how many times the arrow landed on a specific color as compared to the students' predictions. For example, "The arrow never landed on the color we predicted." Or, "We accurately predicted the next color only 1 out of 4 spins."

Describe a change that could be made in this experiment. For instance, what could you do to increase the chances of the arrow landing on any particular color?

We could increase the size of the section of any particular color on the spinner to improve the chances of the arrow landing on it.

What could you do that would make the outcome of the experiment harder to predict?

We could change the sphere to have many unequal sections of different colors.

Unit 2: Building Towers Pg. 14

Describe the outcomes of your probability experience. (Whole class)

Answers should include how many times the spinner landed on each color and how many cubes of like color are in the class tower. For example, the spinner landed on red 20 times and we have 20 red cubes in our class tower.

How are the outcomes connected to what you predicted?

Answers should include how many times the spinner landed on each color as compared to the predictions. For example, we predicted 15 red, but our outcomes were 20 red. Our prediction was close to the actual outcome.

What do you think will happen if you conduct this probability experiment again? Explain your answer.

Answer should reflect an understanding that the size of each tower will correspond in magnitude to the size of the sections of color on the spinner.

Describe a change that could be made in this experiment.

We could change the size of the color sections or the number of sections we will use. For example, we could change the spinner to have three equal sections.

Predict the outcomes of the modified probability experiment. Explain your answer.

If we change the spinner to have three equal sections, it is most likely that it will land on each color one out of three times.

Unit 3: In the Bag! Pg. 19

Describe the outcomes of your probability experience.

Answers should include how many times each statement was true when the crayons were drawn from the bag. For example:

"There are more purple crayons," was true 6 out of 24 draws;

"There are less red crayons," was true 6 out of 24 draws;

"There are exactly two brown crayons," was true 2 of 24 draws;

"At least one of the crayons is red," was true five out of 24 draws; and "There are no yellow crayons," was true five out of 24 draws.

© Teacher Created Resources, Inc.

Answer Key

Page 19, cont.

How are the outcomes connected to what you predicted?

Answers should include how many times each statement was true as compared to the predictions. For example, we predicted:

"There are more purple crayons" would be true 8 out of 24 times. Our outcomes were 6 out of 24. Our prediction was close to the actual outcomes.

What do you think will happen if you conduct this probability experiment again? Explain your answer.

Because the number of colored crayons remains the same, it is likely that the outcomes will be similar, given enough trials (at least 100 draws from the bag of crayons).

Describe a change that could be made in this experiment.

We could change the number of colored crayons in the bag.

Predict the outcomes of the modified probability experiment. Explain your answer.

If we change the number of colored crayons in the bag, one or more of the statements may become more likely. For example, if we place eight purple crayons, two brown crayons, one red crayon, and one yellow crayon in the bag, "There are more purple crayons," will most likely be true.

Unit 4: Chips Away! Pg. 22

Describe the outcomes of your probability experience.

Answers should include how often each score occurred. For example, "I scored 20 five times out of twenty. I can describe my outcomes as a fraction 5/20 = 1/4 or as a percent 25%."

How are the outcomes connected to what you predicted?

Answers should include how many times each score occurred as compared to the predictions. For example, we predicted a score of 20 would happen ten out of twenty times. Our prediction was not as close to the actual outcomes as we wanted.

What do you think will happen if you conduct this probability experiment again? Explain your answer.

Because the numbers on the game board numbers and the number of chips thrown remain the same, it is likely that the results of the next experiment are similar, given enough trials (at least 100).

Describe a change that could be made in this experiment.

We could change the number of chips thrown. For example, we could throw three chips and find the sum to indicate the score.

Predict the outcomes of the modified probability experiment. Explain your answer.

If we use the same game board, but throw more chips, it is likely that the scores will increase in value because there is another addend in the equation.

Unit 5: Digits in a Cup Pg. 26

Describe the outcomes of your probability experience.

Answers should include how many times the digit drawn was more than five, less than five, or exactly five. For example, the number was more than five 10 out of 20 times (1/2 or 50%), less than five 8 out of 20 times (4/10 or 40%), and exactly five 2 out of 20 times (1/10 or 10%). It is more likely that the digit drawn will be less than five because there are five digits less than five (0, 1, 2, 3, 4) and only four digits more than five (6, 7, 8, 9). The least likely outcome is exactly five because the chance of drawing five is one out of ten.

How are the outcomes connected to what you predicted?

Answers should include how many times the digits were drawn as compared to the predictions. For example, we predicted the number five would be drawn two times out of 20. Our prediction was the actual outcome.

#3740 How to ... Data and Probability: Grade 4   46   © Teacher Created Resources, Inc.

# Answer Key

Page 26, cont.

What do you think will happen if you conduct this probability experiment again? Explain your answer.

Because the digits in the cup remain the same, it is likely that the outcomes of the next experiment will be the same given enough trials.

Describe a change that could be made in this experiment.

We could take the zero out leaving only the digits 1-9 in the cup.

Predict the outcomes of the modified probability experiment. Explain your answer.

If we take the zero out of the cup, the probability of drawing less than five and the probability of drawing more than five are now equally likely (each have 4 out of 9 chances). We predict that we will record more than five and less than five the same number of times.

Unit 6: Fill 'er Up   pg. 30

Describe the outcomes of your probability experience.

Answers should include which number cube was used and how many rolls it took to fill the cup. Additionally, students should record how many each number was rolled. For example, we used the first number cube and it took five rolls to fill the cup. We rolled 2, 5, 1, 3, 3. When we used the third cube, it only took three rolls to fill the cup. We rolled 4, 6, 4. Using the second cube, it took eight rolls to fill the cup. We rolled 1, 3, 1, 3, 2, 1, 2, 1.

How are the outcomes connected to what you predicted?

Answers should include how many rolls it took to fill the cup using specific number cubes as compared to the predictions. For example,

"We predicted the third cube would offer a greater chance of filling the cup in the least number of rolls. Our prediction was the actual outcome."

What do you think will happen if you conduct this probability experiment again? Explain your answer.

Because we would use the same number cubes, it is likely that the outcomes of the next experiment will be similar given enough trials.

Describe a change that could be made in this experiment.

We could use a fourth number cube with the digits 7, 7, 8, 8, 9, 9.

Predict the outcomes of the modified probability experiment. Explain your answer.

If we use a fourth cube (with 7, 7, 8, 8, 9, 9), we predict it will have an even greater chance of filling the cup in the least number of rolls because it has larger numbers on it.

Unit 7: Spin to 45   pg. 34

Describe the outcomes of your probability experience.

Answers should include which spinner was used and how many spins it took to reach 45. For example, we used the first spinner (with 1, 2, 3, 4) and it took 18 spins to reach 45. We used the second spinner (with 5, 6, 7, 8) and it only took 7 spins to go over 45.

How are the outcomes connected to what you predicted?

Answers should include how many spins it took to reach 45 using specific spinners as compared to the predictions. For example, we predicted the last spinner (with 1, 0, 0, 10) would offer us the greatest chance to spin 45 because it had a ten on it. Our prediction was not close to the actual outcome. We learned that it was more likely to spin 0 than it was to spin 10 using that spinner.

What do you think will happen if you conduct this probability experiment again? Explain your answer.

Because we would use the same spinner options, it is likely that the outcomes of the next experiment will be the same given enough trials.

Answer Key

Page 34, cont.

Describe a change that could be made in this experiment.

We could use another spinner with 10, 10, 10, 10.

Predict the outcomes of the modified probability experiment. Explain your answer.

If we use another spinner (with 10, 10, 10, 10), we predict it will have an even greater chance of reaching 45 in the least number of spins because it has larger numbers on it.

Unit 8: Make a Square   Pg. 39

Describe the outcomes of your probability experience.

Answers should include which spinner was used and how many spins it took to make a tangram square. For example, we used the first spinner and it took 21 spins to make the tangram square. We used the second spinner and it only took 10 spins to make the tangram square.

How are the outcomes connected to what you predicted?

Answers should include how many spins it took to make the tangram square using specific spinners as compared to the predictions. For example, "We predicted the last spinner (with parallelogram and triangle) would offer us the greatest chance to make a tangram square. Our prediction was not the actual outcome."

What do you think will happen if you conduct this probability experiment again? Explain your answer.

Because we would use the same spinner options, it is likely that the outcomes of the next experiment will be similar given enough trials.

Describe a change that could be made in this experiment.

We could use a spinner divided in half with "triangle" and "any piece."

Predict the outcomes of the modified probability experiment. Explain your answer.

If we use this spinner, we predict it will have an even greater chance of making a tangram square in the least number of spins.

Unit 9: Button Drop for Letters   Pg. 44

Describe the outcomes of your probability experience.

Answers should include how often the button landed on each letter. For example, the button landed on "A" four times out of 100. I can describe my outcomes as a fraction 4/100 or as a percent 4%.

How are the outcomes connected to what you predicted?

Answers should include how many times the button landed on each number as compared to the predictions. For example, we predicted the button would land on "A" 5 times out of 100. Our prediction was close to the actual outcomes.

What do you think will happen if you conduct this probability experiment again? Explain your answer.

Because the letters on the board remain the same, it is likely that the results of the next experiment are similar, given enough trials (at least 100).

Describe a change that could be made in this experiment.

We could change the experiment by using the optional "Button Drop for Letters" chart.

Predict the outcomes of the modified probability experiment. Explain your answer.

If we use the optional chart, it is more likely that the button will land on a vowel because the optional chart includes 17 vowels (a, a, a, e, e, e, i, i. i, o, o, o, u, u, u, u, u) and 11 consonants.